iPhone 17 Pro User Guide

Step-by-Step Manual for Beginners and Seniors to Master iOS 18, Explore Hidden Features, Customize Settings, Improve Battery Life, and Capture Stunning Photos Like a Pro

Georgette Howard

Disclaimer

This book is an independent publication and is not authorized, sponsored, or endorsed by Apple Inc. "iPhone" and "iOS" are trademarks of Apple Inc. Any references to specific features, apps, or updates are for educational purposes only. Every effort has been made to ensure accuracy, but software changes and updates may alter the steps shown.

Preface

A Smarter Way to Learn Your iPhone

Because technology should feel like freedom, not frustration.

There's something beautifully human about the first time you unbox an iPhone. The way the light catches its edges, the gentle hum of possibility when it powers on — a mix of excitement and curiosity that makes us all feel a little like kids again.

But let's be honest — with every new generation, things get more advanced, more automated… and sometimes, a little more confusing. The iPhone 17 Pro and iOS 18 are brilliant, but brilliance should never feel intimidating.

That's why I wrote this book — to give you a smarter, simpler, friendlier way to master your iPhone, without ever feeling lost or overwhelmed.

This isn't another technical manual filled with jargon. It's a real-world, step-by-step iPhone 17 Pro user guide — built to help you see, understand, and enjoy what your iPhone can truly do. Whether you're a first-time user, a senior switching from Android, or someone who's owned iPhones for years but feels left behind by the latest updates, this guide walks beside you — patiently, clearly, and practically.

The Promise of This Book

Inside these pages, you'll learn:

- How to set up your iPhone 17 Pro safely and confidently — from Face ID to Wi-Fi and Apple ID.
- How to navigate iOS 18's new interface, including its redesigned Control Center, widgets, and personalization tools.

- Camera secrets that transform ordinary photos into cinematic memories using Night Mode, ProRAW, and Cinematic video.

- Battery-saving habits and charging tips to extend your iPhone's lifespan — one of the most searched features for iPhone users worldwide.

- Troubleshooting and maintenance skills that can fix 90% of common issues before you ever need a technician.

- Practical steps to secure your privacy, organize your apps, and even use your iPhone as a productivity and wellness companion.

This book covers everything you wish Apple explained more clearly — from hidden iOS 18 gestures to iCloud backup, Apple Wallet setup, FaceTime SharePlay, Siri shortcuts, and digital mindfulness for balanced screen time.

More Than Just a Manual

This is a book for real people — the ones who just want their iPhone to work beautifully.

You'll find:

- Plain-English explanations (no tech dictionary needed).
- Real-life examples that make complex steps easy.
- Illustrations and pictorial guides that visually connect what you read to what you see.
- Smart habits that make your iPhone safer, faster, and more enjoyable.

And because the iPhone 17 Pro is more than a device — it's a digital companion — this book helps you unlock every layer of it: from AI-powered Siri integrations and smart home connectivity to health tracking, wallet management, and mindful use.

Why This Book Stands Out

Search the web and you'll find countless "iPhone 17 Pro

manuals," but few written with care for the human behind the screen.

This guide was created for:

- Seniors and beginners who want confidence with every tap.
- Everyday users who want to use their iPhone to plan, create, and stay organized.
- Tech-curious learners eager to explore hidden features, shortcuts, and camera tricks.
- Busy professionals and creators who need to simplify their digital lives.

It's the perfect blend of clarity, depth, and discoverability — designed to rank high on searches like:

iPhone 17 Pro user manual, iPhone 17 beginner guide, iOS 18 hidden tips, battery optimization, FaceTime and camera tricks, and troubleshooting for seniors.

A Smarter, Friendlier Way Forward

Technology should never make you feel small. It should make you feel capable. The iPhone 17 Pro was designed to empower, not overwhelm — and my goal is to make sure it does exactly that.

So take your time. Turn the pages. Try the steps. Each chapter was written like a personal session — not with a technician, but with a patient friend who wants you to enjoy learning again.

By the end, you won't just know your iPhone — you'll own it. Confidently. Calmly. Completely.

Welcome to the smarter way to learn your iPhone.

Welcome to a future *where technology feels human again*.

Introduction

The iPhone 17 Pro: Where Simplicity Meets Possibility

For more than a decade, the iPhone has done something extraordinary — it has changed not just how we communicate, but how we live. From our first photo to our daily reminders, from morning alarms to midnight playlists, this little rectangle of glass and light has become a trusted companion — one that travels everywhere with us.

And yet, with each new model and update, the iPhone also becomes more complex.

The iPhone 17 Pro, powered by iOS 18, is Apple's most intelligent and capable device yet — but for many users, that intelligence can feel like a maze. Menus have shifted, icons have evolved, and even simple gestures have new meanings.

That's where this book comes in.

The iPhone 17 Pro User Guide was written to help you bridge the gap between technology and everyday life — to make your iPhone simple again.

Whether you're a first-time user, a senior learning at your own pace, or a long-time iPhone fan who just wants to understand the latest features better, this is your map through the world of iOS 18 — written in plain, calm, human language.

What You'll Discover Inside

Each chapter is a complete journey — practical, visual, and empowering.

You'll start with the essentials — setting up your iPhone, understanding the new interface, and securing it with Face ID and passcodes. From there, you'll explore deeper layers of what makes the iPhone 17 Pro truly remarkable.

Here's a glimpse of what's waiting for you inside:

- Clear, step-by-step instructions to set up your device, customize your settings, and connect to Wi-Fi or Bluetooth without confusion.

- A visual guide to iOS 18's redesigned interface — from widgets and Focus Modes to the all-new Control Center and Notification Summary.

- Camera mastery made simple — how to use the Main, Ultra-Wide, and Telephoto lenses, master Cinematic Mode, and shoot stunning photos in Night Mode or ProRAW.

- Everyday solutions for battery drain, overheating, app crashes, and performance lag — practical troubleshooting steps anyone can do.

- Privacy and security settings explained clearly, so your personal data stays yours.

- Productivity tools like Notes, Reminders, Freeform, and Calendar — helping you stay organized and creative.

- Entertainment and wellness apps — from Apple Music and Podcasts to Fitness+ and Screen Time — that balance your digital and real life.

- iCloud backups, Apple Wallet, and Maps — how to use them safely and efficiently.

- Future-ready insights into Apple's direction — including AI-driven Siri, Smart Home integrations, and sustainable technology.

This book turns every tap and swipe into something meaningful — every feature into something usable.

Why This Guide is Different

You won't find overwhelming jargon or technical filler here. Instead, you'll find a human voice that teaches, reassures, and inspires.

Every topic — from FaceTime SharePlay to Shortcuts automation, from camera exposure control to Mindfulness tracking — is written to feel like a one-on-one session with a friendly expert, not a cold manual.

Each section is also supported by visual pictorial explanations, giving you a quick way to see and remember what you're learning. This book was designed to be both read and experienced.

The Human Side of Technology

Apple has always said, *"Technology is most powerful when it empowers everyone."*

That's the heartbeat of this guide. The goal isn't just to teach you how to use your iPhone 17 Pro — it's to help you feel confident, capable, and in control of it.

Because your iPhone isn't just a device — it's a reflection of your rhythm, your creativity, and your life.

When you understand it, everything flows more smoothly.

You communicate more clearly. You capture moments more beautifully. You stay organized, balanced, and safe — not because you're "good with technology," but because your technology has finally become good for you.

A Journey You'll Actually Enjoy

By the time you turn the last page, you'll know your iPhone 17 Pro inside and out — but more importantly, you'll enjoy using it.

You'll stop fearing updates and start welcoming them. You'll start exploring new shortcuts, mastering the camera like a pro, and even helping others around you learn what you now know.

That's the real purpose of this book: to make mastery simple, exploration joyful, and learning effortless.

So, take a deep breath.

You're not behind.

You're right on time — and your iPhone 17 Pro is about

to make life easier, sharper, and more connected than ever.

Welcome to your **complete iPhone 17 Pro guide for iOS 18 — the smarter, simpler way to learn, explore, and thrive with your iPhone.**

Chapter 1

Getting Started with the iPhone 17 Pro

Where first impressions meet lifelong confidence.

There's something almost ceremonial about opening a new iPhone box. The clean, white packaging, the precise way every component fits perfectly in place — it's Apple's subtle reminder that technology can be both beautiful and simple. But beneath that sleek design lies a machine of astonishing capability, waiting for you to bring it to life.

Whether you're upgrading from an older model or holding your first iPhone ever, this chapter will help you begin with ease. From your first tap to your first swipe, you'll see how intuitive the *iPhone 17 Pro* truly is when guided step by step.

Unboxing: What's in the Box and

Why It Matters

When you lift the lid, you'll find your *iPhone 17 Pro* resting like a piece of precision art. Inside, Apple keeps things minimal — your device, a USB-C to Lightning cable, a SIM tray tool (for physical SIM models), and the documentation envelope. You'll notice what's not there: a charging brick. Apple continues its environmental initiative, assuming you already have one compatible charger from a previous device.

Before powering up, take a moment to admire the craftsmanship. The 17 Pro feels lighter yet stronger, with aerospace-grade titanium edges that improve durability without adding bulk. The front Ceramic Shield is now tougher, designed to resist scratches and micro-impacts that used to plague earlier models.

A Quick Tour: Buttons, Ports, and

Design Layout

The physical layout has changed subtly but meaningfully:

- **Right Side:** The elongated **Side button** now doubles as both power and Siri trigger. Press and hold to summon Siri, or quickly tap once to wake or sleep the screen.

- Left Side: You'll find the **volume buttons** and the **Action Button** — Apple's new replacement for the silent switch. This customizable button can toggle silent mode, launch the camera, start voice memos, or activate shortcuts. You decide what it does.

- **Bottom Edge:** The new **USB-C** port replaces the old Lightning port, allowing faster charging and universal cable compatibility.

- **Top Edge:** Sleek and unbroken, home to antenna lines and improved speaker grills for spatial sound.

- Learning the feel of these buttons is like learning the corners of a new home — within days, they'll feel instinctive.

First Boot-Up: Bringing Your iPhone to Life

Press and hold the Side button until you see the iconic Apple logo. As it powers on, you'll be greeted with the welcoming words: "Hello." Swipe up to begin.

The *Quick Start* screen appears, offering two options:

1. **Automatic Setup** – Place your old iPhone near the new one, and it will securely transfer your settings, Wi-Fi, and Apple ID.

2. **Manual Setup** – For new users, tap "Set Up Manually." You'll connect to Wi-Fi, set your region, and follow the onscreen steps.

Choose a language, connect to your Wi-Fi, and continue through the data & privacy screen. Don't rush — the setup process ensures your phone recognizes your unique preferences and security layers from the start.

Understanding Face ID and the Dynamic Island

The front of your iPhone holds two of its most impressive technologies — **Face ID** and the **Dynamic Island**.

Face ID uses an advanced TrueDepth camera system to map your face securely. During setup, slowly move your head in a circular motion as directed. Once registered, you can unlock your phone, authenticate purchases, and log into apps simply by glancing at your screen.

Remember, Face ID adapts to you over time — new glasses, hairstyles, even aging won't throw it off. For added flexibility, you can add an alternate appearance, such as a spouse or family member who may also use the device occasionally.

Above the screen lies Apple's playful innovation: the **Dynamic Island**. Instead of hiding the camera cutout, Apple turned it into an interactive space that changes fluidly — showing incoming calls, Face ID status, charging progress, or background activities like music or timers. Watch closely as it expands, contracts, or splits into two — this "living notch" is the heartbeat of your iPhone's multitasking experience.

Setting Up Your Apple ID and Wi-Fi Safely

Your **Apple ID** is the passport to everything Apple — iCloud, App Store, FaceTime, and backup services. If you already have one, sign in when prompted. If not, create a new one using a secure email address and strong password.

Safety Tip: Avoid using birthdays, names, or repeating numbers in your password. Enable **two-factor authentication** so no one can access your account even if they guess your password.

Next, connect to a **trusted Wi-Fi network**. Tap your network name, enter the password, and wait for the checkmark.

If you're setting up in a public space, avoid open networks — they may expose your information. Once connected, your phone will activate automatically. It might restart during setup; that's normal.

At this stage, iCloud begins syncing — calendars, photos, and backups from your previous devices will start transferring quietly in the background.

Navigating the iOS 18 Home Screen

When the setup completes, you'll land on the **iOS 18 Home Screen** — a refined, living canvas designed for simplicity and focus.

You'll notice:

- **Rounded App Icons:** Minimal and uniform, now adapting to your wallpaper tone for a softer, color-coordinated look.

- **Dynamic Widgets:** Live tiles showing real-time information such as weather, health, and battery. You can move, resize, or stack them by pressing and holding any blank area on the screen until the icons start to jiggle.

- **App Library:** Swipe left past your last Home Screen page to find all your apps neatly organized by category. Use the search bar at the top to locate any app instantly.

- **Control Center:** Swipe down from the top-right corner — a redesigned panel where you control Wi-Fi, Bluetooth, brightness, volume, and your flashlight. In iOS 18, you can rearrange these controls to suit your routine.

Think of this screen as your daily dashboard — everything you need, one glance away.

A Friendly Tip Before You Continue

Before diving into the deeper features, take a few minutes to explore. Tap icons, adjust volume, experiment with wallpaper themes, or simply watch how the Dynamic Island reacts to music and notifications. The best learning often happens through curiosity.

Your iPhone 17 Pro isn't just powerful — it's intuitive. It

learns how you use it and quietly adapts. Over time, it becomes less like a device and more like a digital reflection of you.

Chapter 2

Mastering the New Interface in iOS 18

Your iPhone, reimagined for simplicity and flow.

Every iPhone generation brings a new wave of change —
not just in design, but in how we interact with the world
through our fingertips. iOS 18 takes this philosophy
further, turning the familiar into something smoother,
smarter, and more personal.

For many users, the first few minutes after an update can
feel like walking into a rearranged room — everything's
still there, just in different places. But with a little
guidance, you'll find that the iOS 18 interface is not only
easier to use, but actually anticipates what you want to do
before you even reach for it.

This chapter is your hands-on walkthrough of the
redesigned **Control Center**, **App Library**, **Notification**

Summary, and **Gestures** that make navigation feel second nature. You'll also learn to customize widgets and wallpapers so your iPhone doesn't just work — it feels like yours.

A New Look: Where Elegance Meets Intuition

When you first unlock your iPhone 17 Pro after updating to iOS 18, you'll notice something subtle yet striking: the interface feels calmer. Icons are cleaner, corners are softer, and animations glide instead of snap. Apple refined every pixel to reduce visual fatigue — a small but meaningful nod to users who spend hours a day on their devices.

The focus is on **clarity and movement**. The transitions between screens feel fluid, powered by what Apple calls *"Liquid Glass Motion."* Even system menus respond with gentle depth — so when you swipe up, it feels like you're lifting a layer rather than opening a file.

These small touches are why iOS 18 feels alive, not mechanical.

Understanding the Redesigned Control Center

The Control Center is your iPhone's command hub — where you toggle Wi-Fi, adjust brightness, and switch modes in seconds.

In iOS 18, it has been completely reimagined:

- **Expandable Panels:** Instead of cramped icons, each section expands when tapped — giving quick access to connected devices, volume sliders, or Wi-Fi networks without leaving the screen.

- **Custom Layouts:** You can now rearrange your Control Center tiles. Tap Settings → Control Center

→ Customize Controls and drag to reorder. Want your flashlight or screen recording shortcut at the top? You decide.

- **Smart Stacks:** New grouped tiles automatically adapt based on time or activity — for example, showing Music controls during workouts, or Focus Mode at night.

This flexibility turns your Control Center into something uniquely yours — no clutter, no confusion, just what you need, when you need it.

The App Library: Order Without Effort

Gone are the days of endless swiping through app pages. **The App Library** — introduced a few versions ago and now polished in iOS 18 — is where your apps organize themselves automatically.

Swipe left past your last Home Screen page, and you'll see clean, smart folders labeled Productivity, Creativity, Social, Health, and more. Each folder updates itself, learning from your habits. The apps you use most rise to the top; those you rarely touch fade quietly into the background.

Pro Tip: *Pull down anywhere on the App Library to reveal the **Search Bar** — type a few letters, and your app appears instantly.*

For seniors or beginners, this eliminates the frustration of "I know it's here somewhere." Every app is now two taps away, always organized, always ready.

Customizing Widgets and Wallpapers

Customization in iOS 18 is about self-expression —

subtle, elegant, and meaningful. Your Home Screen is now a living space, not just a grid of icons.

1. **To add a widget:**
   - Touch and hold any blank space on your Home Screen.
   - Tap the "+" icon in the top-left corner.
   - Browse through available widgets — Weather, Calendar, Fitness, Batteries, or even your favorite apps.
   - Choose size (small, medium, large) and drag it where you like.

2. **To customize wallpapers:**
   - Go to Settings → Wallpaper → Add New Wallpaper.
   - Try Depth Effect wallpapers, where the subject subtly overlaps the clock, creating a 3D illusion.

- You can also assign different wallpapers for Focus Modes — perhaps a calm gradient for Work, a vibrant photo for Personal time.

3. **Interactive Widgets:**
   - In iOS 18, widgets are no longer static. You can check off reminders, pause music, or switch devices directly from the widget — no app opening required.

With a few gestures, your phone begins to mirror your personality — your rhythm, your mood, your world.

The New Notification Summary

Notifications can be a blessing or a burden. iOS 18 solves the chaos with **Notification Summary 2.0** — a redesigned system that sorts alerts intelligently.

Instead of bombarding you throughout the day, notifications are now bundled and delivered at chosen intervals, such as morning, afternoon, and evening.

Here's how to set it up:

- Go to Settings → Notifications → Scheduled Summary.

- Toggle on **Scheduled Summary**, then select your preferred times.

- You'll see suggestions for which apps to include — Messages, News, Mail, etc.

The AI-driven ranking system highlights what matters first: calls, messages, and system alerts rise to the top; less urgent updates wait politely below.

For seniors or anyone easily distracted by constant buzzing, this feature restores peace without missing what's important.

Quick Access Gestures and

Shortcuts

Gestures are the hidden language of your iPhone. Once you learn them, everything flows faster.

Here are a few essential ones every user should master:

- **Swipe Up (from bottom edge):** Go to Home Screen.

- **Swipe Up and Hold:** View open apps (App Switcher).

- **Swipe Down (from top-right corner):** Open Control Center.

- **Swipe Down (from middle):** Search your iPhone.

- **Double Tap Back (Settings → Accessibility → Touch → Back Tap):** Assign a custom shortcut — like opening the camera, taking a screenshot, or activating Siri.

In iOS 18, gesture sensitivity has improved dramatically

— smoother transitions, less accidental triggers, and faster response times. It's all about feel. Apple engineers spent months fine-tuning the millisecond difference between "tap" and "press," so your phone behaves intuitively, not impatiently.

A Word About Accessibility

Not every hand moves the same way, and iOS 18 understands that.

If gestures feel tricky, visit *Settings → Accessibility → Touch → AssistiveTouch*. It adds a floating on-screen button that replicates physical gestures with a single tap — ideal for those who prefer easier control.

Other helpful settings include:

- **Reduce Motion:** Minimizes parallax and screen animations.

- **Bold Text & Display Zoom:** Increases readability for better comfort.

- **Haptic Feedback Options:** Adjusts how much vibration you feel with each touch.

Apple's design principle here is simple — technology should adapt to you, not the other way around.

A Final Thought: The Beauty of Simplicity

Once you start customizing and exploring, you'll notice something remarkable: the more personal your iPhone becomes, the simpler it feels.

You're not just learning to use a device — you're shaping an experience.

The *iPhone 17 Pro and iOS 18* were designed for this very harmony: a phone that feels like a natural extension of you.

And as you grow more comfortable, you'll find yourself doing more with less effort — gestures becoming second nature, settings aligning with your habits, and notifications working for you, not against you.

Welcome to the new language of simplicity — fluent, graceful, and unmistakably yours.

The new iOS 18 interface — fluid, personal, and beautifully simple

Chapter 3

Setting Up for Safety and Ease

Because peace of mind is the most powerful feature on your iPhone.

The iPhone 17 Pro isn't just a marvel of design — it's a fortress built for your privacy, safety, and personal comfort.

In a world where data moves faster than thought, Apple's greatest innovation may not be its titanium frame or cinematic camera, but its invisible layers of protection. Your iPhone quietly guards your identity, your habits, and even your eyes — blending technology with empathy.

This chapter will show you how to make your iPhone truly yours: secure, accessible, and tailored to your needs. From Face ID setup to personalized accessibility tools and parental controls, you'll learn how to make your phone not just smarter — but safer, friendlier, and unmistakably human.

Face ID: Your Face Is Your Key

Your iPhone 17 Pro recognizes you — literally.

Face ID is one of Apple's most advanced security systems, using over 30,000 invisible infrared dots to map your unique facial structure. It's faster than typing a passcode and safer than most fingerprint systems.

To set up Face ID:

1. Open **Settings → Face ID & Passcode**.
2. Tap **Set Up Face ID**.
3. Hold your iPhone at arm's length and follow the on-screen circle.
4. Slowly move your head to complete the scan.
5. Tap **Done**.

Once complete, your phone will unlock the moment it sees you — whether you're wearing glasses, a hat, or growing a beard. The TrueDepth system continuously learns your face, adjusting to subtle changes over time.

For added flexibility, you can **add an alternate appearance** (perfect for spouses, caregivers, or trusted family members).

Just remember: Face ID data never leaves your device. It's stored securely in the **Secure Enclave**, not on Apple's servers. Even Apple can't access it.

Passcode Setup: The Unsung Hero of Digital Safety

While Face ID is fast, your **passcode** remains your foundation of security.

Think of it as the master key to your digital home.

To create or change your passcode:

- Go to *Settings* → Face ID & Passcode → Change Passcode.
- Choose a **six-digit** code for strong protection (avoid birthdays or simple sequences).
- For maximum security, select **Custom Alphanumeric Code** and mix letters with numbers.

If your iPhone ever restarts or hasn't been used for over 48 hours, it will ask for this passcode before re-enabling Face ID.

This extra layer ensures that even if your device is lost or stolen, no one can access your personal data without your explicit permission.

Bonus Tip:

*Enable **Erase Data** after 10 failed passcode attempts (Settings → Face ID & Passcode → Erase Data).*

It's your final shield against unauthorized access —

powerful, silent, and reliable.

Accessibility: Making the iPhone Truly for Everyone

Apple's accessibility design is one of the most human aspects of iOS. It's not about limitation — it's about freedom.

Whether your challenge is vision, hearing, motion, or dexterity, iOS 18 is built to meet you halfway, often more.

1. VoiceOver – Your Screen Speaks to You

VoiceOver is Apple's spoken interface for users with visual impairment. It describes everything on your screen — icons, text, even battery level — aloud.

To activate it:

- Go to *Settings → Accessibility → VoiceOver → On.*

- Once active, swipe left or right to move between items, and double-tap to select.

You can adjust the speaking speed and choose a preferred voice. It's intuitive enough that many users without vision loss use it while driving or multitasking.

2. Display Accommodations – Comfort for the Eyes

Bright screens can sometimes be overwhelming, especially at night or for seniors. iOS 18 includes fine-tuned **Display Accommodations** to reduce strain:

- **Reduce White Point:** Lowers the intensity of bright colors.
- **True Tone:** Adjusts the screen to match ambient light for a natural look.
- **Night Shift:** Warms screen tones in the evening to help your eyes relax.
- **Reduce Motion:** Limits screen animations for those sensitive to movement.

You'll find these under Settings → Accessibility → Display & Text Size.

For anyone who uses their iPhone for long hours — these tweaks can mean the difference between fatigue and comfort.

3. Hearing and Touch Adjustments

If you use hearing aids, *Settings → Accessibility → Hearing Devices* lets you pair compatible devices for clear, synchronized audio.

If touch gestures are challenging, you can use **AssistiveTouch** — a floating on-screen button that performs taps, swipes, or screenshots with one press.

Tip: Customize it under Settings → Accessibility → Touch → AssistiveTouch.

It's especially helpful for seniors or users recovering from injury.

Parental Controls and Family Sharing

Technology can be empowering — but for children, it must also be guided. iOS 18 provides **Screen Time and Family Sharing** features that give parents control without invading privacy.

To enable Screen Time controls:

1. Go to *Settings → Screen Time*.

2. Tap **Turn On Screen Time**, then select **This is My Child's iPhone**.

3. Set app limits, downtime hours, and content restrictions.

To create a Family Sharing group:

- Go to *Settings → Your Name → Family Sharing*.

- Add your child's Apple ID to manage purchases, subscriptions, and app permissions.

You can approve or deny app downloads remotely, set age filters for web content, and even track device location securely — all without needing physical access to the device.

Privacy permissions extend to camera, microphone, and location access. iOS 18 now displays a **live indicator dot** whenever these features are used — green for camera, orange for microphone.

It's your silent watchdog for digital transparency.

Privacy Settings That Protect Without Interrupting

Apple's privacy framework is built on a simple belief: Your data is yours.

From email tracking protection to app transparency, your iPhone constantly works behind the scenes to guard what matters.

Here's how to fine-tune it:

- Go to *Settings → Privacy & Security*.

- Review permissions app by app.

- Toggle off access for anything unnecessary (e.g., location for games or camera for unrelated apps).

You can also enable **App Tracking Transparency**, which forces apps to request permission before collecting your data for advertising.

To turn it on:

Settings → Privacy & Security → Tracking → Allow Apps to Request to Track.

When enabled, your digital footprint stays invisible unless you explicitly say otherwise.

Creating an iPhone That Feels Like

Home

Once your phone is secure and accessible, it's time to make it personal.

- Choose a ringtone that feels uniquely "you."
- Set your favorite photo as wallpaper.
- Arrange apps by mood or purpose — one page for creativity, another for calm.

Security and simplicity don't have to feel sterile; your phone should reflect your rhythm, your comfort, and your peace of mind.

By the end of this setup, you won't just have a protected iPhone — you'll have a trusted companion. One that knows when to unlock, when to quiet down, and when to help you focus.

Security meets simplicity – your iPhone, tailored to you.

Chapter 4

Beating Battery Drain for Good

Because power should serve your day, not rule it.

You can tell a lot about an iPhone user by how they treat their battery. Some plug in the moment they see 60%. Others wait until the screen gasps its last red percent. And then there are those who live by superstition — "never charge overnight" or "close all apps or you'll ruin it!"

The truth is, the iPhone 17 Pro has one of the smartest battery systems ever built into a smartphone — but even brilliance needs guidance. This chapter is your blueprint to mastering energy efficiency: how to understand what drains your power, what doesn't, and how to keep your iPhone healthy for years to come.

Understanding Battery Health

Metrics

Every iPhone battery has a story — and the Battery Health section is where it's told.

To view it:

- Go to **Settings** → **Battery** → **Battery Health & Charging**.

Here, you'll see three key readings:

1. **Maximum Capacity (%):**

This represents how much power your battery can hold compared to when it was brand new.

- 100% means your battery is at peak condition.
- Around 80–90% after two years is normal.

2. **Peak Performance Capability:**

This tells you whether your phone can deliver full power when needed (like during gaming or 4K recording). If you ever see "Performance management applied," it means your iPhone has prevented a crash caused by an old battery — a quiet, intelligent safety move.

3. Optimized Battery Charging:

This feature learns your daily routine and slows charging near 100% when you don't need it yet — a protective habit that extends long-term health.

Tip: Always keep "Optimized Battery Charging" on. It's like giving your iPhone a rest during sleep.

The Real Battery Killers: Background Activity

One of the hidden culprits of power loss isn't brightness or Bluetooth — it's background refresh

Many apps continue working silently, fetching updates, checking for messages, or syncing photos even when you're not using them.

To manage this:

- Go to **Settings** → **General** → **Background App Refresh**.
- Choose **Wi-Fi only** or **Off** for most apps.
- Keep it On only for essentials like WhatsApp, Mail, or Health.

You'll be amazed how much longer your battery lasts when your phone isn't multitasking without your permission.

Pro Tip:

Swipe up from the bottom and pause halfway to open the App Switcher. Close apps you haven't used in hours — not obsessively, but strategically.

Over-closing apps can backfire, forcing them to reload from scratch every time. The goal is balance — not control, but cooperation.

Charging Myths vs. Facts

Let's set the record straight on some of the most common myths that quietly drain your battery or peace of mind:

Myth 1: "Charging overnight ruins your battery."

Fact: Not anymore. Modern iPhones stop charging at 80% and use Optimized Charging to reach 100% just before you wake up. Leaving it plugged in overnight is safe — and convenient.

Myth 2: "You must drain your battery to zero before charging."

Fact: Old nickel batteries needed this. Modern lithium-ion batteries prefer frequent top-ups. Try to stay between **20% and 80%** when possible.

Myth 3: "Fast charging shortens lifespan."

Fact: Not if you use Apple-certified chargers. The iPhone 17 Pro manages power intelligently, drawing only what it needs. What harms batteries is heat — not speed.

Myth 4: "Closing all apps saves battery."

Fact: iOS manages background tasks efficiently. Constantly swiping away apps forces them to relaunch, using more energy.

Myth 5: "You shouldn't use your phone while charging."

Fact: Occasional use is fine, especially for light tasks. Just avoid heavy gaming or 4K recording while plugged in, as both increase heat.

Smart Charging Habits for Everyday Users

A few mindful changes can dramatically extend your battery's lifespan:

1. **Avoid Extreme Temperatures:**

Your iPhone loves room temperature (20–25°C). Avoid direct sunlight, hot dashboards, or freezing pockets.

When your phone feels warm, give it a short break — it's not being dramatic, just protective.

2. Use Certified Cables and Chargers:

Cheap chargers can deliver unstable current, causing micro-damage over time. Stick with MFi-certified accessories.

3. Unplug Fully Charged Power Banks:

Leaving your phone connected to a fully charged power bank keeps it cycling between 99% and 100% — subtle but harmful in the long run.

4. Reduce Brightness and Auto-Lock Time:

Go to *Settings* → *Display & Brightness* → *Auto-Lock* and choose **30 seconds or 1 minute**.

Lower brightness manually or enable Auto-Brightness —

small steps, big impact.

5. Turn Off Location Services for Non-Essential Apps:

Location tracking is a silent drainer. Manage it under Settings → Privacy → Location Services.

Turn it off for apps that don't need constant access (e.g., shopping or news apps).

6. Switch to Dark Mode at Night:

On OLED screens like the iPhone 17 Pro, black pixels consume less energy.

Go to Settings → Display & Brightness → Dark Mode → Automatic.

Power Management in Different

Power Contexts

If you live in regions where electricity isn't always stable — or frequent power outages are common — battery management becomes even more crucial.

Inverter or Generator Charging:

When using backup power sources, make sure your inverter provides clean output (Pure Sine Wave). Cheap, unregulated inverters can cause voltage spikes that damage your phone's battery circuit.

If using a generator, wait 1–2 minutes after starting it before plugging in your charger — allowing current to stabilize.

USB Hubs and Car Chargers:

If you rely on car chargers during long commutes, choose models with built-in surge protection. For laptops or USB hubs, avoid chaining too many devices — power delivery weakens with each split.

Tip for African and Asian countries:

Use Surge Protectors and Smart Extension Boxes with voltage control.

They're inexpensive safeguards that can add years to your phone's lifespan.

Low Power Mode: The Everyday Lifesaver

Low Power Mode is Apple's polite way of saying, *"Let's make it last."*

When your battery drops below 20%, you'll see the prompt — but you can enable it anytime.

Go to **Settings** → **Battery** → **Low Power Mode**, or add it to your **Control Center** for one-tap access.

What it does:

- Reduces background activity

- Lowers screen brightness slightly

- Disables Mail fetch and visual effects

The result? You can stretch 15% battery for nearly two extra hours — a lifesaver when traveling or caught without a charger.

Long-Term Battery Care: Think Years, Not Hours

The most overlooked battery advice is the simplest: respect the rhythm of charge and rest.

Every lithium-ion battery has a lifespan of **500–800 full charge cycles** (from 0% to 100%). That means each time you partially charge — say from 40% to 80% — you're preserving a fraction of that cycle instead of using one up.

With balanced habits, your battery can retain **90% capacity even after three years**, saving you the cost of a replacement and the frustration of mid-day charging anxiety.

Your goal isn't perfection — it's consistency. Treat your iPhone's battery the way you'd treat your own energy: recharge often, avoid extremes, and rest when possible.

A Closing Thought: Power as Freedom

A well-managed battery isn't just about longer screen time — it's about freedom. The freedom to capture one more sunset photo, finish a long call with family, or navigate home without watching the red icon blink in panic.

The iPhone 17 Pro gives you control; your habits give it endurance. Together, they form a simple truth: *smart energy is peace of mind.*

Power smarter, not harder.

Chapter 5

Solving Overheating and Performance Lag

Keep your iPhone cool, calm, and consistently powerful.

It starts as a small warmth in your hand — a little heat behind the camera, a glow near the battery — and before long, your iPhone feels uncomfortably hot, and performance begins to stumble. Apps stutter. The camera lags. Battery drains faster than you expect.

Every iPhone, no matter how advanced, produces heat. What matters isn't that it gets warm — it's why and how you respond. The iPhone 17 Pro, with its titanium body and A18 chip, is designed to balance performance and temperature intelligently. But when heat persists, or your phone feels unusually slow, it's your cue to look closer.

This chapter helps you understand the difference between harmless warmth and real trouble — and shows how to restore your iPhone's steady, efficient rhythm.

Why Your iPhone Heats Up

Heat is a byproduct of work. Just like a car engine or a laptop fan, your iPhone generates warmth when it's doing more — processing, rendering, charging, or updating.

Common everyday triggers include:

- Long FaceTime or video calls
- Continuous camera or 4K recording
- Gaming or augmented reality apps
- GPS navigation under direct sunlight
- Streaming or downloading large files
- Charging while in a tight pocket or case

Most of the time, this warmth is **normal** and temporary. The iPhone is designed to cool itself internally by regulating processor speed and brightness. But if the heat feels persistent, even when idle, or affects performance, there's more to uncover.

Hardware Heat vs. Software Heat

Knowing the difference helps you pinpoint the cause quickly.

Hardware Heat

This comes from physical stress — components working hard or external conditions pushing limits.

- **Charging Heat:** Slight warmth while charging is normal, especially when battery levels are low. But excessive heat can indicate poor ventilation or an overused charger.

- **Direct Sunlight:** Leaving your iPhone on a car dashboard or near a window under the sun can cause it to display the "Temperature: iPhone needs to cool down" warning.

- **Cases & Covers:** Thick or poorly ventilated cases can trap heat, especially leather or plastic ones without airflow.

When you feel heat at the back or edges during charging or under sunlight, it's hardware-related. Simply cool it naturally — avoid fridges or fans; ambient cooling is safest.

Software Heat

This happens when apps or background processes overwork the CPU or GPU.

Typical culprits:

- Rogue apps consuming power in the background
- Unfinished downloads or updates
- Widget overloads or misbehaving extensions
- iCloud or Photos syncing large media batches

Software heat often appears after major iOS updates or app installations — the system needs time to re-index data, optimize storage, or sync files. If your phone heats up after an update, give it 24–48 hours to stabilize before troubleshooting further.

How to Check What's Causing the Heat

Apple gives you simple tools to diagnose the issue without third-party apps.

1. **Go to Settings → Battery.**
   - Scroll to "Battery Usage by App."
   - Look for any app showing unusual percentage usage or background activity.
2. **Go to Settings → General → iPhone Storage.**
   - Large "Documents & Data" numbers under a single app (like Safari or Facebook) can signal runaway storage buildup.
3. **Force Quit or Reinstall Problem Apps.**

- Swipe up from the bottom and pause halfway to open the App Switcher.

- Close apps not in use, especially those running location services or camera access.

4. **Restart your iPhone.**

- A simple reboot clears memory clutter and resets processor tasks — a digital "cool bath."

Managing Widgets and Rogue Apps

Widgets are helpful — but every live tile, animation, or background refresh requires processing power.

In iOS 18, interactive widgets (music, weather, fitness) can occasionally over-refresh, especially if you've added too many on one Home Screen.

Here's how to optimize them:

- Step 1: Touch and hold a widget until you see the "Edit Widget" option.

- Step 2: Disable "Live Update" for widgets you don't need updating every few seconds.

- Step 3: Use Smart Stacks — they rotate widgets intelligently based on time and usage, saving memory.

Next, tackle **rogue apps** — those consuming resources in the background even when closed.

To manage them:

- Go to *Settings* → *General* → *Background App Refresh*.

- Turn it off entirely or select *Wi-Fi only*.

- For power-hungry apps like TikTok, Snapchat, or Meta apps, disable background activity completely.

You'll notice your iPhone runs cooler and smoother — often within a day.

Cooling Tips That Actually Work

If your iPhone feels hot, resist the temptation to panic or overreact. Here's how to cool it safely:

1. Remove the Case:

Let your phone breathe. Metal and glass radiate heat faster when exposed to air.

2. Stop Charging Temporarily:

Charging while hot compounds the issue. Unplug until it cools.

3. Turn On Airplane Mode:

This disables network tasks temporarily, helping your phone rest.

4. Dim the Brightness:

Go to Control Center → Brightness Slider. The display is a major heat source — lowering it works wonders.

5. Avoid Cold Surfaces or Refrigeration:

Sudden temperature changes can cause condensation inside components — damaging your iPhone permanently.

6. Close Heavy Apps:

Shut down games, camera apps, or editing tools until temperature stabilizes.

Extra Tip:

If your iPhone overheats during calls or videos, switch to speaker mode or wireless earbuds — holding it against your skin transfers additional warmth.

When to Seek Service

Some overheating patterns indicate deeper issues that require professional attention.

Contact Apple Support or visit an authorized service center if you notice:

- Constant overheating even in standby mode

- Battery swelling or screen lifting (a sign of battery damage)

- Rapid draining (over 10% drop within minutes of light use)

- Frequent shutdowns with "Temperature" alerts

- Device heating only near one corner or port area

Apple's diagnostics tools can run thermal scans and check for software loops or failing power regulators — issues no home fix can resolve.

Restoring Performance After Overheating

When your phone overheats, it automatically throttles performance — meaning apps may lag or animations feel sluggish. Once the temperature normalizes, speed returns on its own.

To help it recover faster:

1. Restart after Cooling: Clears temporary throttling commands.

2. Check for Updates: Apple often releases background patches for heat and battery optimization.

3. Reset Settings (if needed): Go to Settings → General → Transfer or Reset iPhone → Reset All Settings. This won't delete data but refreshes configuration.

Your phone will feel fresh, responsive, and balanced again.

Preventing Future Heat & Lag

- Keep iOS updated regularly.
- Use official chargers and avoid power banks that overheat.

- Don't stack your iPhone on laptops or routers — they radiate heat too.

- Keep your storage below 90% capacity for smoother performance.

- Restart weekly to clear caches and background threads.

These habits don't just prevent overheating — they preserve the youth of your iPhone, keeping it fast and stable well into future iOS updates.

A Final Thought: Cool Is the New Fast

Technology mirrors us more than we realize — we too perform best when calm, rested, and balanced.

Your iPhone 17 Pro is no different. With smart habits, it runs cooler, lives longer, and gives you consistent performance without compromise.

Remember: overheating isn't a flaw — it's feedback. It's your device saying, *"I'm doing too much."*

The fix is almost always simple: breathe, simplify, optimize — and let your iPhone find its calm.

Keep it cool —
performance follows calm.

Chapter 6

The Camera Masterclass: Shoot Like a Pro

Your lens, your story, your power.

Photography is the art of noticing — and your iPhone 17 Pro gives you the tools to notice beautifully.

Behind its sleek titanium body lies one of the most powerful camera systems ever placed in a smartphone. Three lenses, powered by intelligent software and Apple's new A18 Pro chip, work together like a small orchestra — each one performing a different role in perfect harmony.

You don't need to be a professional to capture breathtaking shots. What you need is understanding — of light, of angles, of how each lens sees the world. That's where this chapter becomes your personal photography mentor.

Understanding the Triple-Lens

System

The iPhone 17 Pro's camera system combines three distinct perspectives, each designed for a different storytelling angle.

1. The Main Lens (1x – Your Everyday Eye)

This is your go-to lens — the one you'll use 70% of the time.

It captures crisp, balanced images with a natural field of view, similar to what your eyes see. Perfect for portraits, family moments, food shots, and everyday life.

With a wider aperture and improved sensor stabilization, it thrives in both bright sunlight and gentle evening glow. Even handheld, it captures detail without blur.

When to use it:

- General photography
- Portraits with natural background blur

- Indoor shots where lighting varies

Tip: Step closer instead of zooming. Physical proximity adds emotion, sharpness, and intimacy to your photos.

2. The Ultra-Wide Lens (0.5x – Your World in Full View)

Imagine being able to fit an entire landscape, building, or crowd into one frame without stepping back. That's the Ultra-Wide lens.

It sees nearly twice as much as the Main lens, perfect for travel photography, architecture, and creative perspectives.

When to use it:

- Landscapes and cityscapes
- Group photos in tight spaces
- Dramatic "from-the-ground" angles

Pro Tip: Keep the horizon level. Ultra-wide distortion can bend edges; tilt too much, and straight lines curve unnaturally.

Use it creatively — a low-angle ultra-wide shot of a beach can make the world feel vast and cinematic.

3. The Telephoto Lens (3x or 5x – Your Distance Vision)

The Telephoto lens is your magnifying glass — for moments you can't physically approach. It brings subjects closer without losing detail, maintaining beautiful compression and background separation.

When to use it:

- Portraits with depth and bokeh
- Wildlife or street photography
- Concerts, events, or architecture details

Pro Tip: *The telephoto lens works best in good lighting. In dim environments, your iPhone may automatically switch to digital zoom on the main lens. To stay sharp, tap the "1x" label manually to toggle back.*

The Magic of Modes: Cinematic, Night, and ProRAW

Cinematic Mode — Your Personal Film Studio

The Cinematic Mode turns your iPhone into a pocket filmmaker's dream. It automatically shifts focus between subjects, creating depth-of-field effects once possible only on professional cameras.

To activate: Open Camera → Swipe to "Cinematic."

You can tap any face or object to change focus. As the subject moves, your iPhone intelligently follows, keeping them sharp while gently blurring the background.

Pro Tip:

After shooting, open the video in the Photos app → Tap Edit → Adjust focus points frame by frame. It's like directing your own movie after the scene has been shot.

Night Mode — Capturing Light in the Dark

Night Mode activates automatically in low-light environments, lengthening exposure time to let in more light. You'll see a small moon icon appear.

How to use it best:

- Hold your iPhone steady — use both hands or prop it on a surface.
- Watch the timer (1–3 seconds typical, longer in very dark scenes).
- Move subjects slowly or keep them still.

You'll get results that defy logic — starlit skies, glowing streets, candlelight dinners — all beautifully detailed without grain.

Pro Tip: The best Night Mode photos come when you embrace light sources — lamps, candles, or even passing cars add creative texture.

ProRAW — The Photographer's Playground

ProRAW is Apple's format for professionals who crave control. It captures all the raw image data before processing, giving you flexibility for editing exposure, shadows, highlights, and tone.

To enable ProRAW:

Settings → Camera → Formats → Turn on Apple ProRAW.

When to use it:

- Landscapes with high contrast
- Creative edits in apps like Lightroom or Snapseed
- Professional shoots needing color correction

ProRAW photos are larger files, so use them selectively — when artistic precision matters.

Framing, Lighting, and Composition

Tips

Even the smartest camera can't fix poor composition —
but a few simple habits can elevate your photography from
casual to captivating.

1. Follow the Rule of Thirds

Enable the grid under Settings → Camera → Grid.

Imagine your frame divided into nine equal rectangles.
Place your subject at the intersections — not the center.
This creates natural balance and draws the viewer's eye.

2. Chase Good Light

Light is the soul of photography. The golden hours —
early morning or late afternoon — paint everything in soft,
flattering warmth.

Avoid harsh midday sun unless you want strong shadows.
Indoors, face your subject toward a window or use
reflective surfaces to bounce light.

3. Use Leading Lines

Roads, railings, or hallways guide the viewer's gaze straight to your subject. Compose intentionally — lines should invite the eye inward, not lead it away.

4. Mind the Background

A cluttered background steals attention. Move a few steps left or right, crouch slightly, or use Portrait Mode to blur distractions.

5. Capture Emotion, Not Perfection

The best photos feel alive. A laugh, a gust of wind, or a glance caught mid-motion often tell more story than a technically flawless pose.

Editing Photos and Videos Directly in the Photos App

Your iPhone 17 Pro gives you professional-grade editing

tools without needing extra apps.

Basic Edits

1. Open your photo → Tap Edit.

2. Use sliders for **Exposure**, **Brilliance**, **Highlights**, and **Shadows**.

3. Adjust **Warmth** or **Tint** for natural skin tones.

Filters and Crop

- Use Apple's subtle filters like Vivid or Dramatic Warm for mood enhancement.

- Straighten horizons or crop tighter to strengthen focus.

Video Editing

You can trim, apply filters, adjust brightness, or even stabilize shaky clips directly in Photos.

Open → Edit → Adjust tab → Drag sliders for light, contrast, or color balance.

Pro Tip:

Duplicate the original before heavy editing (Tap Share →
Duplicate → Edit Copy) so you can always revert if
needed.

Bonus: The "Photographer's Mindset"

Remember — the camera doesn't make the photograph; you do.

When you lift your iPhone, don't just ask, "What do I see?" Ask, "What do I want others to feel?"

That single shift — from capturing to storytelling — transforms every shot you take.

Photography is less about perfection and more about presence. You don't chase light; you learn to wait for it. You don't take pictures; you make them. And when you start seeing like that, even ordinary moments — coffee steam, a smile, a quiet street — become extraordinary.

Your lens, your world — captured with purpose.

Chapter 7

Customizing Your Experience

Because your iPhone should feel as unique as you.

Every iPhone starts the same — same icons, same wallpaper, same grid of possibilities. But it doesn't stay that way for long. The moment you make it yours — rearrange the icons, choose a wallpaper that calms or inspires you, set a Focus mode for your work or your rest — that's when your iPhone becomes more than a device. It becomes a mirror of your rhythm, your style, your life.

The iPhone 17 Pro paired with iOS 18 brings personalization to its peak — blending design, function, and emotion. Whether you crave simplicity or controlled chaos, this chapter will help you craft a setup that feels perfectly in sync with your world.

Changing Icons, Wallpapers, and Widgets

Personalization starts with visuals — the look and feel of your screen. It's how your phone greets you every time you unlock it.

1. Changing App Icons

Gone are the days when you had to live with standard app designs. The Shortcuts app allows you to give every app its own identity.

To create custom icons:

1. Open the **Shortcuts app**.
2. Tap "+" → **Add Action** → **Open App**.
3. Choose your desired app (e.g., Messages, Mail).
4. Tap the blue icon → **Add to Home Screen**.
5. Tap the small app icon → **Choose Photo or File**.

Now, pick a custom image from your camera roll or one downloaded from an icon pack. Choose something that matches your mood or theme — minimal black and white, pastel tones, or neon gradients.

You can even organize icons by color or category for aesthetic harmony. For example, group all creative apps in warm tones and productivity apps in cool tones.

Tip: Visit Settings → Home Screen → App Library Only if you prefer a minimalist look — letting your custom icons take the spotlight.

2. Transforming Your Wallpaper

Wallpapers are more than decoration — they set your emotional tone. The iPhone 17 Pro's Super Retina XDR display makes every background feel alive.

To set or change a wallpaper:

- Go to *Settings → Wallpaper → Add New Wallpaper.*

- Choose from categories like **Photo Shuffle, Gradient, Astronomy**, or **Emoji**.

Photo Shuffle lets you display different photos throughout the day — morning sunrises, family smiles, evening skies.

Gradient wallpapers match the system theme, shifting subtly between colors depending on light or dark mode.

Pro Tip: *Use a wallpaper that contrasts with your icons — bright wallpaper for dark icons or vice versa — to keep everything easy on the eyes.*

3. Widgets That Work the Way You Do

Widgets in iOS 18 are not just for display; they're alive. They update in real-time and now let you interact directly — toggle music, mark tasks done, or view your calendar without opening an app.

To add or edit widgets:

1. Touch and hold a blank space on your Home Screen until apps jiggle.

2. Tap "+" in the upper-left corner.

3. Choose a widget (Weather, Fitness, Calendar, etc.).

4. Select size and drag it into place.

You can even **stack widgets** — drag one on top of another to create a Smart Stack that changes dynamically throughout your day.

For example:

- **Morning:** Calendar and Weather.
- **Afternoon:** Fitness and Reminders.
- **Evening:** Music or Podcasts.

It's like having your day visualized on your Home Screen — effortlessly.

Setting Focus Modes and Do Not

Disturb

We live in a world of constant notifications. But peace, clarity, and focus are still possible — thanks to Apple's **Focus Mode**.

Focus Modes are an evolution of Do Not Disturb, allowing you to silence distractions based on what you're doing or where you are.

To create or edit a Focus Mode:

1. Go to *Settings → Focus*.

2. Choose a preset like **Work**, **Sleep**, **Personal**, or tap "+" to create your own.

3. Select which apps and contacts can reach you during that time.

4. Customize the Home Screen — hide non-work apps while working, or silence Slack notifications during dinner.

Each Focus mode can have its own **lock screen**, **wallpape**r, and **status message** (e.g., "Notifications are silenced").

Pro Tip: *Link Focus Modes to Time or Location.*

For example:

- **Work Focus:** Turns on at 9 AM, silences social apps, activates professional wallpaper.
- **Personal Focus:** Turns on after 6 PM, hides work apps, activates music and family widgets.
- **Sleep Focus:** Automatically reduces screen brightness and notifications after 10 PM.

This feature transforms your phone from a constant interrupter into a mindful companion.

Do Not Disturb — The Classic Shield

While Focus is smart, sometimes you just want a single

switch that silences everything.

Swipe down from the top-right corner to open *Control Center*, then tap the **moon icon**.

That's **Do Not Disturb** — simple, effective, immediate.

It mutes calls, alerts, and notifications but allows alarms and emergency contacts through if configured.

When you need space — a nap, prayer, or pure silence — this is your instant calm button.

Automations and the Shortcuts App: Making Life Effortless

Automation is where your iPhone starts thinking with you, not just for you.

The **Shortcuts app** lets you chain actions together — little routines that happen automatically based on triggers you define.

Here are a few examples that can transform your day:

Morning Routine

- **Trigger:** 7:00 AM or when your first alarm stops.
- **Actions:** Play your favorite playlist → Read today's weather → Open Calendar → Announce "Good Morning" via Siri.

Focus Routine

- **Trigger:** When Work Focus activates.
- **Actions:** Turn on Low Power Mode → Set brightness to 60% → Open Notes → Enable Do Not Disturb.

Night Routine

- **Trigger:** 10:00 PM.
- **Actions:** Lower screen brightness → Play soft music → Turn on Sleep Focus → Enable Night Shift → Set alarm.

To set one up:

1. Open **Shortcuts** → **Automation** → + → **Create Personal Automation**.

2. Choose your trigger (time, app opened, Wi-Fi connection, etc.).

3. Add desired actions from the menu.

Pro Tip: *Tap **Ask Before Running** → **Off** to make it fully automatic.*

Shortcut Ideas That Feel Like Magic

Here are a few that most iPhone 17 Pro owners love:

- **Battery Saver:** When battery drops below 20%, automatically turn off Bluetooth, lower brightness, and enable Low Power Mode.

- **Arrival Mode:** When connecting to home Wi-Fi, turn on Wi-Fi lights, open Spotify, and disable Do Not Disturb.

- **Drive Mode:** When connecting to your car Bluetooth, start navigation to work and send an "On my way" text.

Each of these saves seconds — but together, they save energy, focus, and thought. The goal isn't just efficiency; it's freedom from digital friction.

Your iPhone, Your Flow

When your phone is truly personalized, it doesn't distract — it blends. It becomes an assistant that anticipates you, adapts to you, and grows with you.

The point of all this customization isn't to make your iPhone look pretty — it's to make it feel right.

When you unlock it, it should feel like stepping into a familiar rhythm: everything where it should be, nothing more, nothing less.

Your iPhone 17 Pro is not just a piece of technology — it's a reflection of you. And like you, it's always evolving.

Designed by you — for the way you live.

Chapter 8

iCloud, Backups, and Storage Made Simple

Because your memories deserve to be safe.

There's a quiet kind of comfort in knowing that no matter what happens to your iPhone — whether it's lost, damaged, or replaced — your photos, messages, and memories will always find their way back to you.

That peace of mind lives in one simple word: **iCloud**.

iCloud isn't just a storage service; it's the invisible thread connecting all your Apple devices — your iPhone, iPad, Mac, even your Apple Watch. It remembers so you don't have to. This chapter will help you understand how to use it confidently, how to manage space wisely, and how to back up your entire digital life without stress.

Understanding iCloud — Your

Digital Safety Net

Think of iCloud as your personal digital vault — always on, always secure, and accessible anywhere.

It automatically syncs your photos, notes, contacts, calendar events, and app data across every Apple device signed in with your Apple ID. That means when you take a picture on your iPhone, it appears on your iPad or Mac within seconds — no cables, no effort.

To set up iCloud:

- Go to *Settings → [Your Name] → iCloud*.
- Sign in with your Apple ID (or create one if you haven't yet).
- Toggle on the apps and services you want to sync — Photos, Contacts, Messages, Notes, and so on.

From that moment on, every change you make is mirrored safely in iCloud. Delete a contact, it disappears everywhere. Add a note, it shows up instantly across all your devices.

Pro Tip: *Keep "Find My iPhone" turned on under iCloud → Find My — it's not just for lost devices. It helps you locate, lock, or even erase your phone remotely if needed.*

iCloud Drive — More Than Backup

While iCloud keeps your apps and settings safe, iCloud Drive is where you can store, organize, and share actual files — documents, PDFs, and even entire folders.

To access it:

- Go to the **Files app** on your iPhone.
- Tap **Browse → iCloud Drive**.

You can create new folders, move files, or share items via links — all updated instantly across your devices.

It's especially powerful for students, writers, and professionals who switch between devices often. Work on a document at home on your Mac, then continue editing it later on your iPhone during a commute — seamlessly.

Pro Tip:

Enable Desktop & Documents Folders under Settings → iCloud Drive → Options if you use a Mac. It automatically syncs your workspaces across all your devices.

Managing Storage Efficiently

By default, Apple gives you **5 GB of free iCloud storage**. It's enough to get started, but as your photos, videos, and backups grow, you'll likely need more.

To check your current usage:

Go to *Settings → [Your Name] → iCloud → Manage*

104

Storage.

Here, you'll see a color-coded bar showing how much space each category (Photos, Backups, Messages, etc.) uses.

You have three main ways to manage this:

Option 1: Optimize Storage

- Go to *Settings → Photos → Optimize iPhone Storage.*

  This keeps smaller, space-saving versions of your photos on your device while full-resolution originals stay safely in iCloud.

  When you open a photo, it downloads instantly — saving gigabytes of space.

Option 2: Clean Up Backups

- Old or unused backups from previous devices can eat space.

Go to *Settings* → *iCloud* → *Manage Storage* → *Backups* and delete outdated ones you no longer need.

Option 3: Upgrade Your iCloud Plan

- Apple offers flexible plans:

  - 50 GB

  - 200 GB (perfect for families)

  - 2 TB (ideal for creators and photographers)

Plans are affordable and can be shared via **Family Sharing**, meaning everyone in your family has their own private space under one plan.

Backing Up Your iPhone — Your Digital Insurance

Your iCloud backup is what saves your phone's soul — the apps, settings, photos, and conversations that make it yours.

To enable automatic iCloud Backup:

1. Go to *Settings → [Your Name] → iCloud → iCloud Backup*.

2. Turn on **Back Up This iPhone**.

3. Plug in your phone and connect to Wi-Fi — iCloud will back up automatically when idle and charging.

You can also tap **Back Up Now** to trigger it manually anytime.

What gets backed up:

- Photos and videos (if iCloud Photos is off)

- Device settings

- App data and home screen layout

- Messages and call logs

- Health and HomeKit data

If your iPhone is ever lost or replaced, simply sign in with your Apple ID on the new device and choose "Restore from iCloud Backup." Within minutes, your digital life reassembles itself — like magic.

Restoring from a Backup

If you get a new iPhone or need to reset your device, restoring is simple:

1. Power on the new iPhone.
2. Follow the setup instructions until you reach the Apps & Data screen.
3. Choose **Restore from iCloud Backup**.
4. Sign in and select your most recent backup.

Your settings, messages, and apps return exactly as you left them — even your wallpaper and home screen layout reappear. It's like stepping back into your old phone, but fresher.

Offline Backups: The Alternative

(Mac or PC)

If you prefer physical control over your data, you can back up manually to a computer using Finder (Mac) or iTunes (Windows).

To back up manually:

- Connect your iPhone via cable.

- Open Finder or iTunes.

- Select your device, then click **Back Up Now**.

- For maximum security, enable **Encrypt Local Backup** — this includes passwords, Wi-Fi settings, and Health data.

Pro Tip: Combine both — keep one iCloud backup (for convenience) and one offline backup (for redundancy). Professionals call this the 3-2-1 rule:

3 total copies of your data → 2 different types of storage → 1 stored off-site (iCloud).

iCloud Shared Albums — Memories Meant to Be Shared

Sharing memories doesn't have to mean sending hundreds of photos manually. With iCloud Shared Albums, you can create collaborative galleries that family and friends can access, add to, and comment on.

To create one:

1. Open the **Photos app** → **Albums** → + → **New Shared Album.**
2. Name it (e.g., "Family Vacation 2025") and invite people.
3. Add photos, and others can contribute their own.

It's like a private social space for your favorite people — no ads, no data mining, just shared memories.

Privacy and Security: Your Data,

Your Control

iCloud is designed with privacy at its core. Apple uses **end-to-end encryption** for sensitive data like passwords, Health, and Messages — meaning even Apple cannot access your content.

For ultimate protection, turn on **Advanced Data Protection** (Settings → iCloud → Advanced Data Protection → On). This encrypts your iCloud backups and Photos, ensuring only your devices can decrypt them.

Always use **two-factor authentication** for your Apple ID — your most powerful shield against unauthorized access.

A Final Thought: Simplicity with Security

The beauty of iCloud is that it does the hard work quietly.

You don't have to think about it. You just live your life, capture moments, and stay connected — while iCloud keeps everything safe in the background.

Your iPhone 17 Pro may be brilliant, but it's the unseen safety of iCloud that makes it fearless. So whether it's your baby's first photo, your college project, or a lifetime of notes, remember this simple truth: **your memories deserve protection, not luck**.

Your digital life–safely stored, always within reach

Chapter 9

Communication Essentials

Staying connected — clearly, calmly, and in control.

At its heart, the iPhone has always been about connection — not just the speed of a message or the clarity of a call, but the human pulse behind every "hello." The iPhone 17 Pro refines this even further. Whether you're texting, calling, or video chatting, it's not just communication — it's presence.

This chapter walks you through the tools that make connection effortless: Messages that feel personal, FaceTime that feels real, and Mail that feels organized. You'll also learn how to manage noise — the spam calls, unknown numbers, and distractions that clutter modern communication.

Mastering Messages — Where Conversations Live

Messages in iOS 18 go beyond simple texting. They're smarter, faster, and more expressive — designed to help you communicate your way, not just type words.

1. Cleaner Messaging for a Clearer Mind

If your Messages app feels crowded, start fresh with these small habits:

- **Pin Important Conversations:** Swipe right on a chat → tap Pin. Your favorite people now stay at the top.
- **Use Search Wisely:** Swipe down in the Messages app — type a name, photo keyword, or file type to find anything instantly.

- **Group Threads Simplified:** Name your group chats (tap the top header → Change Name and Photo) for quick recognition.
- **Hide Alerts for Busy Hours:** Swipe left on a conversation → tap the bell icon to mute it temporarily.

Your inbox will start to feel less like a flood and more like a living space — curated and peaceful.

2. Express More with Fewer Words

Apple added subtle but powerful upgrades in iOS 18 to make texting feel natural:

- **Tapback Reactions:** Press and hold a message to react with emojis like ♥, 👍, or 😄.
- **Inline Replies:** Tap and hold a specific message → Reply. Perfect for busy group chats where context matters.

- **Edit and Undo Messages:** Sent something too soon? You have 15 minutes to fix it — simply press and hold the message → Edit or Undo Send.

- **Text Effects:** Type your message, then press and hold the send arrow → choose Slam, Gentle, Invisible Ink, or Loud to match your tone.

And if you're a visual communicator, explore Stickers, Memojis, and GIFs — they add warmth where words might fall short.

FaceTime — Conversations That Feel Real

FaceTime on the iPhone 17 Pro is now a world-class communication tool — no longer just for video chats, but for shared experiences. The A18 chip enhances video quality, while iOS 18 introduces better lighting and background blur that automatically adjust to your environment.

1. FaceTime SharePlay — Watch, Listen, and Learn Together

SharePlay lets you enjoy content with others during a call — movies, music, or even fitness workouts.

To start:

1. Begin a FaceTime call.

2. Open an app like Apple TV, Music, or Fitness.

3. Tap **SharePlay → Share My Screen or Content**.

Everyone sees or hears the same thing, perfectly synced. Great for remote family nights, study sessions, or music jam circles.

Pro Tip: *Use AirPods during SharePlay for crisp, echo-free audio.*

2. Group FaceTime — Up to 32 People, Seamlessly

No more crowding around screens. You can now host

group calls with up to 32 participants.

- Open FaceTime → New FaceTime → Add Contacts.
- When someone speaks, their tile automatically enlarges.

You can switch between **Grid View** (to see everyone equally) and **Speaker View** (to highlight who's talking).

For work calls: You can even share your screen to present documents, slides, or websites in real time.

3. Live Captions — Accessibility That Includes Everyone

With iOS 18, Live Captions automatically generate text from spoken words during calls — perfect for users with hearing loss or for noisy environments.

To enable:

- Go to *Settings → Accessibility → Live Captions (Beta)*.

Once on, you'll see real-time subtitles appear on your screen during FaceTime calls, videos, or even live audio.

This feature embodies Apple's ethos: inclusion through design.

Mail — Order in the Inbox

Apple's Mail app has matured into a powerful yet calm workspace — prioritizing clarity and efficiency over clutter.

1. Focused Inbox

Mail now sorts important messages automatically — newsletters and promotions move aside, while personal or professional emails rise to the top.

Use **VIP Mailboxes** to keep messages from key people in

one place.

- Open an email from a contact → Tap their name → Select **Add to VIP**.

2. Schedule, Snooze, and Undo

- **Schedule Send:** Tap and hold the send arrow → choose "Send Later" to time your email perfectly.
- **Snooze Emails:** Swipe right → Tap Snooze to delay reading until later.
- **Undo Send:** After sending, tap "Undo" at the bottom within 10 seconds.

3. Smart Attachments

When you type "I've attached" but forget to add one, Mail gently reminds you before sending. It's small design details like this that save embarrassment and time.

Pro Tip:

If you use multiple accounts (work, personal, or school),

assign color labels in Settings → Mail → Accounts → Label Color. It helps visually separate inboxes.

Managing Spam Calls and Unknown Numbers

Unwanted calls and messages can steal both time and peace. Thankfully, iOS 18 offers quiet, powerful ways to filter the noise.

1. Silence Unknown Callers

This feature sends unknown numbers straight to voicemail while still showing them in recent calls.

- Go to *Settings → Phone → Silence Unknown Callers → On.*

You'll still receive calls from your contacts, recent calls, and Siri Suggestions — everyone else goes quietly to voicemail.

2. Block Persistent Numbers

If a number keeps bothering you:

- Open *Phone → Recents → "i" → Block this Caller.*

You can manage your blocked list under Settings → Phone → Blocked Contacts.

3. Filter Unknown Messages

Avoid text scams and junk messages by separating them from your main inbox.

Go to Settings → Messages → Unknown & Spam → Filter Unknown Senders → On.

Messages from strangers move into a separate "Unknown Senders" tab — out of sight, but still accessible if needed.

4. Report Junk

If you receive a spam text, scroll to the bottom and tap Report Junk. Apple automatically sends metadata (not your message) to help identify spam patterns.

Pro Tip: Focus Mode for Communication

To truly master balance, link your communication habits to Focus Modes (explained in Chapter 7).

For example:

- **Work Focus:** Allow only Slack, Email, and Family messages.
- **Personal Focus:** Limit notifications to Messages and FaceTime.
- **Sleep Focus:** Silence everything except emergency contacts.

This transforms communication from chaotic to intentional — your iPhone serves your peace instead of demanding your attention.

A Final Word: Connection with Intention

Connection is power — but mindful connection is peace.

Your iPhone 17 Pro can keep you in touch with everyone, everywhere — but it can also help you stay present with the few who truly matter.

Learning how to message, call, and filter with purpose turns your device from a noise machine into a clarity companion. And in that quiet balance, you rediscover what technology is really for — not endless contact, but meaningful connection.

Stay connected,
not consumed.

Chapter 10

Everyday Productivity: Turning Your iPhone Into a Life Tool

Where organization meets inspiration.

For most people, an iPhone is a communication device. But for those who know how to use it wisely, it's something far more powerful — a digital life assistant.

Your iPhone 17 Pro can plan your day, store your ideas, keep your finances tidy, remind you of deadlines, and even help you collaborate with others across the world. And it does it all quietly, elegantly, and efficiently.

This chapter will show you how to make your iPhone a genuine productivity tool — not by adding more apps, but by mastering the ones already built into it.

Notes: Your Second Brain

Think of the Notes app as the most reliable personal assistant you'll ever have — one that never loses a page and always stays in sync.

1. Capturing Ideas Instantly

To open Notes faster, swipe down from the top-right corner → tap the **Control Center Notes icon**.

You can also tell Siri: "Take a note." She'll instantly record your thought hands-free.

Pro Tip: *Pin important notes (swipe right on a note →
tap the pin) so they always appear at the top.*

2. Organizing Notes for Clarity

In iOS 18, Notes supports folders, subfolders, and tagging:

- Use **#tags** within notes to group ideas ("#recipes," "#meetingnotes").

- Long-press a note → move it into a specific folder.

- Tap the new Smart Folder option to auto-organize all notes with a certain tag.

3. Turn Notes Into Documents

Scan papers or receipts right inside Notes:

Open a note → tap **Camera icon** → **Scan Documents**.

You can sign them digitally using your finger or Apple Pencil (if connected).

Now your random ideas, paperwork, and journal entries all live neatly in one searchable place — your personal "cloud brain."

Reminders: Never Miss What Matters

The **Reminders app** isn't just for lists — it's your personal accountability coach.

You can set reminders for time, location, or even context.

1. Time and Location-Based Reminders

- "Remind me to buy milk when I leave home."

- "Remind me to send the report at 9 AM."

Your iPhone will do exactly that — automatically.

Go to *Reminders* → *New Reminder* → *Info (i)* → *Set Date/Time or Location.*

2. Organize by List and Tag

Create separate lists for Work, Family, Groceries, or Fitness.

Use **color-coded icons** for instant recognition.

Tap **Tags** to group similar tasks across lists.

3. Smart Lists

In iOS 18, Apple added Smart Lists that automatically sort tasks due today, flagged items, or completed ones — saving you the mental energy of managing your to-do list.

Calendar: Command Your Time, Don't Chase It

The **Calendar app** is your control center for everything time-related.

The goal isn't to fill every hour — it's to protect the important ones.

1. Master Your Week at a Glance

Switch between **Day, Week**, and **Month** views easily by pinching in or out.

Long-press on a time slot to create a new event instantly.

2. Integrate Multiple Calendars

If you use Google Calendar, Outlook, or a work account — go to *Settings* → *Calendar* → *Accounts* → *Add Account.*

Now you'll see all events in one unified view, color-coded for clarity.

3. Add Alerts and Travel Time

Never be late again:

Tap an event → *Edit* → *Add Travel Time.*

Your iPhone calculates how long it'll take to get there and reminds you when to leave, factoring in traffic conditions via Maps.

Pro Tip: *Sync your Focus Modes with your calendar (Chapter 7). When "Work Focus" activates, your Calendar automatically surfaces only relevant events.*

Freeform: Collaboration Without

Boundaries

Freeform is Apple's visual workspace — part whiteboard, part brainstorming wall, part project planner.

It's designed for creative thinkers, students, teams, and anyone who loves to map ideas visually.

How to Start:

- Open the **Freeform app.**
- Tap the "+" to create a new board.
- Use sticky notes, text boxes, arrows, or sketches to arrange your ideas.

You can invite others to join your board using Messages or FaceTime — and everyone sees changes live.

Perfect for:

- Project planning with colleagues
- Designing study maps
- Visual journaling

- Event or trip planning with friends

Everything syncs across devices automatically via iCloud, so your ideas travel with you.

Wallet, Maps, and Apple Pay — Everyday Convenience

Your iPhone 17 Pro is also your wallet, navigation tool, and travel companion — all in one.

1. Wallet — Carry Less, Do More

Wallet isn't just for cards anymore. You can now store:

- Boarding passes
- Movie tickets
- Transit cards
- Hotel keys
- Event passes

To add a card or pass:

Go to Wallet → + → Add Card or Pass → Follow prompts.

When near supported terminals, just hold your iPhone near the reader — Face ID does the rest.

2. Apple Pay — Fast, Secure, Contactless

Set up once, and you'll rarely touch your wallet again.

Setup:

- Go to *Settings → Wallet & Apple Pay → Add Card.*

- Follow the verification process from your bank.

At checkout, double-press the **Side button**, glance at your screen for Face ID, and tap your iPhone to pay.

Why it's safer:

Apple Pay doesn't share your actual card number — it uses an encrypted token, keeping your data private.

Bonus Tip:

You can now track order receipts and delivery updates directly inside Wallet — perfect for online shoppers.

3. Maps — Navigate Smarter, Travel Better

The latest **Apple Maps** isn't just about directions — it's about context.

It now includes detailed 3D landmarks, improved walking routes, and Offline Maps, so you can navigate even without signal.

Pro Tips:

- **Add Favorite Locations:** Pin your home, office, and frequent destinations.
- **Use Look Around:** Explore neighborhoods with interactive 3D previews.
- **Transit Mode:** Track live bus and train schedules in supported regions.
- **Weather Layers:** Tap the weather icon to check temperature and rainfall along your route.

Apple Maps now integrates with Siri — say, *"Take me to the nearest pharmacy,"* and she'll not only guide you but show estimated arrival times and traffic alerts.

Creating Daily Flow: Integrating It All

Once you've mastered Notes, Reminders, Calendar, and Freeform, your iPhone stops being a device — it becomes a *daily rhythm*.

Here's how they work together:

- Draft ideas in **Notes**.
- Turn those ideas into actionable **Reminders.**
- Schedule them into **Calendar**.
- Visualize plans or collaborations in **Freeform**.
- Pay, travel, and move through your day with **Wallet**, **Maps**, and **Apple Pay**.

Each one complements the other — like pieces of a quiet, efficient ecosystem built around your life, not the other way around.

Bonus: The Art of Digital Balance

True productivity isn't about doing more — it's about doing what matters most, with less friction.

The iPhone 17 Pro is designed to reduce noise, automate tasks, and create time for creativity. Use its power wisely.

At its best, your iPhone should not demand your attention — it should free it.

Plan, create, and live –
all from one device.

Chapter 11

Entertainment, Health, and Mindfulness

Your iPhone isn't just smart — it's also soulful.

There's a quiet joy in realizing your iPhone can do more than connect you — it can center you. It can help you unwind after a long day, build better health habits, and restore balance when life feels overwhelming.

The iPhone 17 Pro, with iOS 18, is designed not only to entertain but to nurture your well-being — blending technology, creativity, and mindfulness in ways that feel deeply human.

This chapter is your guide to that harmony — how to use Music, Podcasts, and Fitness+ not as distractions, but as tools for inspiration and growth.

The Soundtrack of Your Life: Apple

Music & Podcasts

Music is emotion set to rhythm. Podcasts are wisdom set to words. Together, they turn your iPhone into a portal for creativity, motivation, and reflection.

Apple Music — Listen Intentionally

Apple Music is more than a library of songs — it's a universe of moods, curated playlists, and sonic journeys.

To start:

1. Open **Music → Browse → For You.**
2. Explore curated playlists like *"Chill Mix,"* *"Workout Hits,"* or *"Morning Motivation."*
3. Tap the + icon to add favorites to your library.

You can create **custom playlists** that reflect your mood or schedule —

- *Morning Boost:* upbeat tracks to start your day
- *Focus Flow:* instrumental tracks for work

- *Evening Calm:* soft jazz or acoustic for unwinding

Pro Tip:

*Use Siri **Shortcuts:***

"Hey Siri, play my focus mix."

"Hey Siri, shuffle evening playlist."

Apple Music now integrates **crossfade** and **adaptive EQ**, so transitions are seamless — like a continuous emotional wave.

Podcasts — Learn While You Live

Podcasts turn idle moments — a walk, a commute, even cooking — into windows of learning.

To explore:

1. Open the **Podcasts** app → Tap Browse.
2. Check "Top Charts" or "Categories" (Health, History, Science, Self-Improvement).
3. Tap **Follow** to get automatic updates.

Smart listening tip:

Speed up playback (1.25x or 1.5x) to cover more while staying engaged.

Download episodes for offline listening when traveling or during quiet time.

Try these inspiring categories:

- **Motivation & Mindset:** "The Daily Stoic," "On Purpose with Jay Shetty."

- **Wellness & Health:** "Huberman Lab," "Feel Better, Live More."

- **Creativity & Culture:** "99% Invisible," "The Tim Ferriss Show."

Listening isn't escape — it's enrichment.

Fitness+: Move with Meaning

In an age of screens and schedules, movement is medicine

— and **Apple Fitness**+ brings that medicine home.

To start:

- Open the **Fitness app** → **Fitness+ tab.**
- Choose from categories like **Yoga, HIIT, Core, Dance, Strength, or Mindful Cooldown**.
- Select **Trainer** or **Duration** (5–45 minutes).

Apple Fitness+ syncs with your **Apple Watch** (if you have one), displaying heart rate and calories in real time. No watch? You can still follow along — progress tracks via your iPhone's motion sensors.

Pro Tip:

Fitness+ offers Collections — curated programs like "Perfect Your Posture" or "Beginner Strength."

Follow these to build structure and see measurable improvement.

Motivation Hack:

Use **Music Integration.** Every Fitness+ class has handpicked Apple Music playlists that amplify energy. One beat at the right moment can shift your mood — and your mindset.

Apple Health — Your Wellness Dashboard

Health isn't just steps or calories; it's data that tells the story of your lifestyle.

The **Health app** transforms that data into insights that matter.

1. Setting Up Health Tracking

Open **Health** → **Summary** → **Favorites**.

Choose what to display — Steps, Sleep, Mindfulness Minutes, Heart Rate, or Nutrition.

You can link other apps and devices (Fitbit, smart scales,

sleep monitors) under *Health* → *Sharing* → *Apps* → *Add Data Sources*.

2. Mindful Metrics

Track trends like:

- **Sleep Duration:** Consistency improves focus and immune function.
- **Heart Rate Variability (HRV):** Measures stress and recovery balance.
- **Screen Time:** Correlates mental energy with daily phone use.

3. Share Health Data Securely

You can share selected health stats with a doctor or loved one.

Go to *Health* → *Sharing* → *Share With Someone* → *Choose Data Types*.

Apple encrypts everything — your data remains private

and protected.

Mindful Use and Digital Detox

Even the smartest device can become overwhelming if you never step away. The key to mindful living isn't rejecting technology — it's reframing it. Your iPhone can help you disconnect without guilt.

Screen Time — Awareness That Empowers

Go to *Settings* → *Screen Time* to see your daily and weekly usage reports.

It shows how often you pick up your phone, which apps consume time, and when you're most active.

From there, set limits gently — not as punishment, but as boundaries.

- **App Limits:** Restrict social apps after 9 PM.
- **Downtime:** Schedule phone-free evenings.

- **Always Allowed:** Keep essentials like Calls or Music accessible.

Pro Tip:

*Enable **Focus Filters** (linked to your Focus Modes). For example, when Work Focus activates, your iPhone automatically hides personal app notifications.*

It's not about deprivation — it's about reclaiming focus.

Mindfulness in a Digital World

iOS 18 deepens its Mindfulness features in the Health app.

You can log moods daily with color gradients — from calm to stressed — and reflect briefly on what influenced them. Over time, it forms a "mental map" of your triggers and positive patterns.

To practice mindfulness:

Open **Health** → **Mindfulness** → **Start a Session**.

147

Choose **Breathe, Reflect**, or **Focus**.

Follow soft animations that guide your breathing — each inhale and exhale slows your rhythm, grounding you back into stillness.

Bonus: Use **Wind Down Mode** (Settings → Focus → Sleep).

It dims your screen, plays soothing sounds, and gently transitions you to rest — the digital version of tucking yourself in.

When Entertainment Becomes Enrichment

Your iPhone can entertain, educate, and restore you all in the same breath.

A playlist can lift your spirit. A podcast can shift your mindset. A guided workout can rebuild confidence. A moment of silence can reset your entire day.

Technology isn't the enemy of peace — unawareness is. When you use your iPhone intentionally, it becomes an ally in your mental and physical wellness journey.

The iPhone 17 Pro isn't just a device; it's a partner in your growth — if you let it be.

Balance your digital and real life

Chapter 12

Troubleshooting & Maintenance

Confidence begins where confusion ends.

Every piece of technology — no matter how advanced — will have moments of hesitation, just like we do. An app may freeze. A Wi-Fi connection may drop. Your screen might lag, or your iPhone may feel slower than usual.

But here's the truth: 90% of issues don't need a technician. They simply need a little understanding.

This chapter will show you how to keep your iPhone 17 Pro healthy, responsive, and reliable for years — and when it's time to let Apple's experts step in.

1. When Things Slow Down: Understanding Performance

Hiccups

If your iPhone feels sluggish or unresponsive, it doesn't mean it's "old" — it usually means it's working overtime.

Apps, caches, background updates, and large media files can quietly build up and strain your system.

Here's what to do:

Step 1: Restart — the Digital Reset

Press and hold the **Side button + Volume Up (or Down)** until you see the *slide to power off screen*.

Wait a few seconds, then power back on.

It's simple, but powerful — restarting clears temporary memory, reboots background processes, and often solves app lag or battery drain instantly.

Step 2: Clear Background Clutter

- Swipe up from the bottom and pause halfway → swipe up each app to close.

- Go to *Settings* → *General* → *Background App Refresh* → *Off (or Wi-Fi only)*.

Step 3: Free Up Storage Space

A nearly full iPhone runs slower because it lacks breathing room.

Go to *Settings* → *General* → *iPhone Storage*.

Delete large apps, old downloads, or duplicated photos.

Pro Tip: *Enable Offload Unused Apps to remove rarely used apps but keep your data safe.*

2. Connectivity Errors — The Silent Disruptors

When Wi-Fi, Bluetooth, or cellular data misbehave, it often feels like your iPhone has a mind of its own. The fix, however, is usually simple.

Wi-Fi Issues

- Toggle **Wi-Fi off** → **wait 10 seconds** → **turn it back on.**
- Forget and reconnect: Settings → Wi-Fi → (i) next to network → Forget This Network → Reconnect.
- Restart your router — half the time, it's the real culprit.

If still stuck:

Reset network settings: *Settings → General → Transfer or Reset iPhone → Reset → Reset Network Settings*.

(You'll need to re-enter Wi-Fi passwords afterward.)

Bluetooth Problems

- If your AirPods or accessories won't connect:
- Turn Bluetooth off/on.

- Forget the device, then re-pair.

- Check for interference (metal desks, microwaves, or multiple paired devices).

Mobile Data Dropouts

- Go to *Settings → Cellular → Cellular Data Options → Enable 5G Auto.*

- Ensure *Low Data Mode* is off.

- Toggle *Airplane Mode* for 10 seconds to refresh the connection.

Pro Tip: *Carriers in some regions auto-switch between 4G and 5G bands. If you experience drops, manually switch to 4G LTE for more stable connectivity.*

3. App Crashes and Misbehavior

Every app can stumble — even Apple's own. The key is not panic, but patience.

Quick Fixes:

- **Force quit** the app and reopen it.

- **Update** it via App Store → Profile Icon → Update All.

- **Reinstall it:** Delete the app → Re-download → Log back in.

If multiple apps misbehave, check for a system update (see below).

4. Updating iOS — Your Built-In Repair Kit

Updates don't just add features — they quietly fix bugs, improve speed, and patch security gaps.

Always keep your iPhone 17 Pro up to date.

To check for updates:

Go to Settings → General → Software Update →

Download and Install.

Turn on **Automatic Updates** for peace of mind.

Your iPhone will install them overnight when plugged in and connected to Wi-Fi.

Pro Tip: *If your update seems stuck, make sure you have at least 50% battery and 5 GB of free space. Restarting before updating also helps smooth installation.*

5. System Cache & Safari Clean-Up

Over time, apps and websites store temporary files to speed things up — but too much of it can actually slow you down.

Clear Safari Cache:

Go to *Settings → Safari → Clear History and Website Data.*

Offload App Cache:

Most apps (like Instagram or TikTok) store cache that builds quickly. You can't clear it manually, but deleting and reinstalling the app resets its storage.

Reclaim System Cache:

1. If your phone feels bloated overall:

2. Restart.

3. Delete temporary downloads and large attachments in Messages → Info → Documents & Data.

Keep your iOS Storage bar under 80% for ideal performance.

6. Battery and Overheating Checks

You already learned to "beat battery drain" in Chapter 4, but for maintenance:

- Go to *Settings* → *Battery* → *Battery Health & Charging.*

- If Maximum Capacity drops below 80%, consider a battery service.

- Use Optimized Charging to slow wear.

- Avoid third-party chargers — they may degrade battery chemistry over time.

If your iPhone gets hot:

- Stop charging temporarily.

- Close high-power apps (camera, games).

- Keep it out of direct sunlight.

When it cools, resume normal use.

7. Resetting Without Losing Everything

Sometimes you just need a clean slate — without deleting your data.

Option 1: Reset All Settings

(Settings → General → Transfer or Reset iPhone →
Reset → Reset All Settings)

This restores factory defaults for settings only — your photos and apps remain untouched.

Option 2: Full Factory Reset

Only do this if nothing else works or you're selling the device.

- *Settings → General → Transfer or Reset iPhone → Erase All Content and Settings.*
- Ensure your data is backed up to iCloud before proceeding.

After reset, you can restore everything from your iCloud backup (see Chapter 8).

8. When to Contact Apple Support

Despite your best efforts, some issues are simply beyond

home repair — and that's perfectly okay.

You should reach out to Apple Support when:

- Your iPhone doesn't power on or charge.

- You notice battery swelling or screen lifting.

- Face ID or camera modules malfunction.

- Persistent overheating, freezing, or touch unresponsiveness.

- Water damage indicators turn red inside SIM tray (avoid DIY drying attempts).

How to Contact Support:

- Open *Apple* Support app → Choose your device → Follow troubleshooting prompts or schedule a Genius Bar appointment.

- Alternatively, visit support.apple.com/iphone

.If under warranty or AppleCare+, service is often free or discounted.

9. Proactive Care — Keep It Running Like New

Like any good tool, your iPhone benefits from small, regular maintenance habits:

- Restart weekly.

- Update monthly.

- Clean your screen gently with a microfiber cloth.

- Keep ports dust-free using a soft brush (never pins or metal).

- Review storage quarterly.

- Protect it with a quality case, tempered glass, and surge-safe charger.

These rituals keep your iPhone 17 Pro fast, safe, and dependable — ready for every update and adventure ahead.

10. A Mindset Shift: Prevention Over Panic

Technology ages gracefully when cared for.

The difference between a frustrating phone and a flawless one isn't luck — it's attention.

When you understand your iPhone's language — the small signals, the subtle slowdowns — you begin to notice how it mirrors your habits.

Take care of it, and it will take care of you.

Because maintenance isn't about repair; it's about respect — for your time, your memories, and your peace of mind.

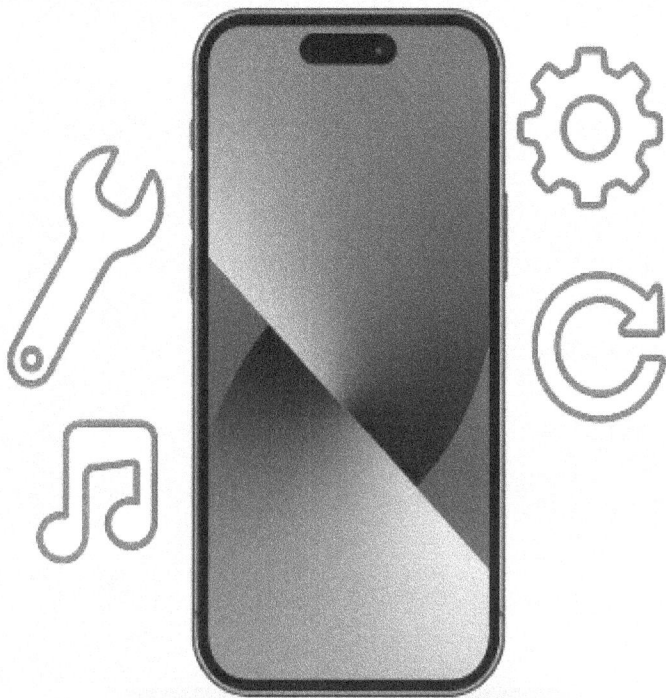

Care for your device,
and it will care for you.

Chapter 13

The Future of iOS: What's Next

Because innovation never truly ends — it evolves.

Every year, Apple does something subtle yet extraordinary: it doesn't just upgrade its devices; it reimagines the way we live with them.

The iPhone 17 Pro is already a marvel of design, intelligence, and efficiency, but what's most exciting isn't what it is — it's what it's *becoming*.

The next wave of iOS development will move beyond features and apps. It will move toward understanding — technology that doesn't just respond but anticipates, learns, and cares for the world around it.

Let's look at three frontiers that define where Apple — and your iPhone — are headed.

1. AI Integration in Siri: From Assistant to Companion

Siri's next chapter isn't about doing more — it's about doing better.

For years, Siri has been the voice of your iPhone. Soon, it will also become its mind. Powered by advanced on-device AI, Apple's vision for Siri moves from simple commands to true contextual understanding.

Conversational Intelligence

Instead of reacting to isolated requests, Siri will soon hold meaningful conversations — understanding tone, intent, and continuity.

Ask:

"*Remind me to call Mom when I get home.*"

and later say,

"*Actually, make it tomorrow morning,*"

and Siri will adapt instantly without needing clarification.

Predictive Awareness

Imagine Siri knowing you're driving and automatically silencing non-urgent notifications, or suggesting playlists that match your current energy level.

Or offering proactive help:

"You usually order groceries on Thursdays — would you like me to open your shopping list?"

That's the future: **AI that feels intuitive, not intrusive**.

Privacy-First AI

Unlike cloud-based assistants, Apple's approach will rely on on-device intelligence. That means your data — your voice, patterns, and habits — stays yours. The goal isn't to collect, but to protect.

The Shift: From an assistant you command → to a companion that understands.

2. Smart Home Connectivity: Living in Sync

We've entered the age of the intelligent home, where your devices don't just coexist — they collaborate.

With upcoming iOS and HomeKit expansions, Apple aims to make smart homes feel less like technology and more like orchestration.

Unified Home Intelligence

Currently, your iPhone can control your thermostat, lights, locks, and appliances via the Home app. Soon, it will *coordinate* them automatically:

- Lights dim when you start a movie.
- Air conditioning adjusts based on your heart rate or room occupancy.
- Your morning alarm triggers coffee brewing, music playback, and gradual curtain lift — all seamlessly.

Cross-Platform Harmony

Through standards like **Matter**, Apple is expanding compatibility with non-Apple devices — meaning your iPhone could soon control almost every connected gadget, from Samsung refrigerators to Philips bulbs, through one ecosystem.

Pro Tip: Keep your Home app updated — Apple is slowly introducing features that allow shared automations and more intuitive voice control via Siri.

Home as a Health Hub

Future Apple Homes may also integrate environmental sensors — tracking air quality, humidity, and even allergens.

Your iPhone could alert you:

"Your bedroom air quality dropped slightly — would you like to open the window or activate the purifier?"

Smart living, in the truest sense, will mean *living well*.

3. Sustainability Features: The Greener iPhone Vision

As technology advances, Apple's greatest innovation may

not be in performance — but in conscience.

Apple has already pledged to be **carbon-neutral across its entire product line by 2030**. The iPhone 17 Pro's titanium frame and recycled materials were only the beginning.

Eco-Design Philosophy

Future devices will use more renewable aluminum, recycled rare-earth metals, and eco-friendly glass.

But sustainability doesn't stop at hardware — it's built into iOS itself.

- **Energy-Saving Algorithms:** Adaptive power management will adjust performance dynamically, extending battery lifespan while reducing environmental strain.

- **Eco-Charging Mode:** Expected soon — a feature that prioritizes renewable energy hours in your region before charging fully.

- **Repairability:** Apple's Self Service Repair program will likely expand, letting users replace batteries and screens more easily without voiding warranties.

Digital Sustainability

Apple is also rethinking software longevity. The next iOS generations are expected to maintain compatibility with older models for longer — reducing electronic waste by design.

In the near future, your iPhone may even recommend donation or repurposing options when you upgrade.

Because sustainability isn't just about materials — it's about *mindset*.

4. What It Means for You

The future of iOS isn't distant — it's unfolding with every quiet update, every background optimization, every new "beta" that refines how you interact with your digital world.

But at its heart, the goal remains the same: **technology that amplifies life, not overshadows it**.

In the coming years, you'll see your iPhone evolve from a tool into an ecosystem of empathy — one that understands your habits, supports your goals, and contributes to a world that values sustainability as much as innovation.

This is Apple's true genius: it never just builds devices. It builds relationships — between people, technology, and the planet we share.

A Glimpse Beyond

Imagine this:

You wake up, and your iPhone — integrated with Apple Health, Maps, and Home — senses your mood, sets your lights to soft amber, starts your favorite playlist, and shows you a mindful quote of the day.

Outside, solar-powered servers sync your data securely, while Siri quietly adjusts your calendar to avoid back-to-back meetings.

That's not science fiction — it's the trajectory of iOS.

A future where intelligence feels gentle, automation feels human, and innovation feels responsible.

A Final Reflection

Every Apple device begins with the same promise: to make life better.

And as iOS continues to evolve, that promise remains the north star — reminding us that progress isn't just faster chips or sharper screens.

It's a world where your technology doesn't just work — it understands.

The iPhone 17 Pro is your gateway to that world — a

future that's already unfolding, right in your hand.

Tomorrow's iPhone —
smarter, kinder, greener.

Quick Reference Cheat Sheet — Gestures & Shortcuts

Your pocket guide for navigating the iPhone 17 Pro with ease.

Essential Gestures

- **Swipe Up (from bottom edge):** Go Home.

- **Swipe Up and Pause:** View open apps (App Switcher).

- **Swipe Down (top-right corner):** Open Control Center (brightness, Wi-Fi, battery).

- **Swipe Down (middle of screen):** Spotlight Search.

- **Swipe Left / Right on Home Screen:** Move between pages or Today View.

- **Pinch to Zoom:** Photos, Maps, and Safari.

- **Double Tap (back of device)** (if enabled via Accessibility → Touch → Back Tap): Custom shortcut (e.g., screenshot or app launch).

Power & Emergency

- **Side Button + Volume Up/Down:** Access power-off and emergency SOS.

- **Hold Side Button (long):** Activate Siri manually.

Quick Actions

- **Press and Hold App Icon:** Open contextual shortcuts (e.g., new note, recent call).

- **Three-Finger Swipe Left:** Undo typing in Notes or Messages.

- **Three-Finger Tap:** Redo typing.

- **Pinch with Three Fingers:** Copy selected text.

- **Spread Three Fingers Apart:** Paste text.

Screenshot & Screen Recording

- **Side Button + Volume Up:** Take a screenshot.

- **Control Center → Record icon:** Start screen recording (tap red bar to stop).

Keyboard Tricks

- **Hold Spacebar:** Turn keyboard into a touchpad for cursor movement.

- **Double Space:** Inserts a period.

- **Tap and Hold Globe icon:** Switch between languages or emoji keyboards.

Time Savers

- **Swipe Left on Notification:** Manage or clear it.

- **Press and Hold Link (Safari):** Open in new tab or add to Reading List.

- **Control Center → Long Press Icons:** Reveal hidden toggles like AirDrop or Personal Hotspot.

Master these, and your iPhone becomes pure instinct — every action just a flick, tap, or glance away.

Glossary of iPhone Terms

AirDrop:

A fast, wireless file-sharing feature for Apple devices using Wi-Fi and Bluetooth.

App Library:

The last page on your Home Screen — automatically organizes apps by category.

Apple ID:

Your personal login for Apple services (App Store, iCloud, iMessage, etc.).

Face ID:

Apple's facial recognition system for unlocking, authentication, and Apple Pay.

Focus Mode:

A customizable setting that filters notifications based on your activity (Work, Sleep, Personal).

Haptic Touch:

Tactile feedback from the screen when you press and hold to reveal extra options.

iCloud:

Apple's online service that syncs and stores your photos, files, and backups securely.

Live Photos:

A short 3-second video captured around your still image, bringing moments to life.

ProRAW / ProRes:

Professional photo and video formats that retain high-quality editing data for creators.

Safari:

Apple's built-in web browser for fast, private browsing.

Siri:

Apple's voice assistant — capable of answering questions,

controlling settings, and automating tasks.

Spotlight Search:

A universal search tool for apps, contacts, web results, or quick calculations (swipe down on Home Screen).

Widgets:

Live tiles that show real-time info like weather, calendar events, or reminders.

Acknowledgement

To every aspiring photographer and filmmaker who dares to pick up a camera and tell a story, this book is for you. Special thanks to my family and friends for their encouragement, and to the creative community whose passion inspires me daily. Your support made this guide possible.

www.ingramcontent.com/pod-product-compliance
Lightning Source LLC
Chambersburg PA
CBHW081815200326
41597CB00023B/4254